Light
We Can See

Illustrations: Janet Moneymaker
Design/Editing: Marjie Bassler

Light We Can See
ISBN 978-1-950415-28-1

Published by Gravitas Publications Inc.
Imprint: Real Science-4-Kids
www.gravitaspublications.com
www.realscience4kids.com

When you wake up in the morning, you might see the Sun shining.

Time to get up!

The Sun is a big star that sends out **light.** The light from the Sun that we can see is called **visible light**. **Visible** means that something can be seen with our eyes.

I see you!

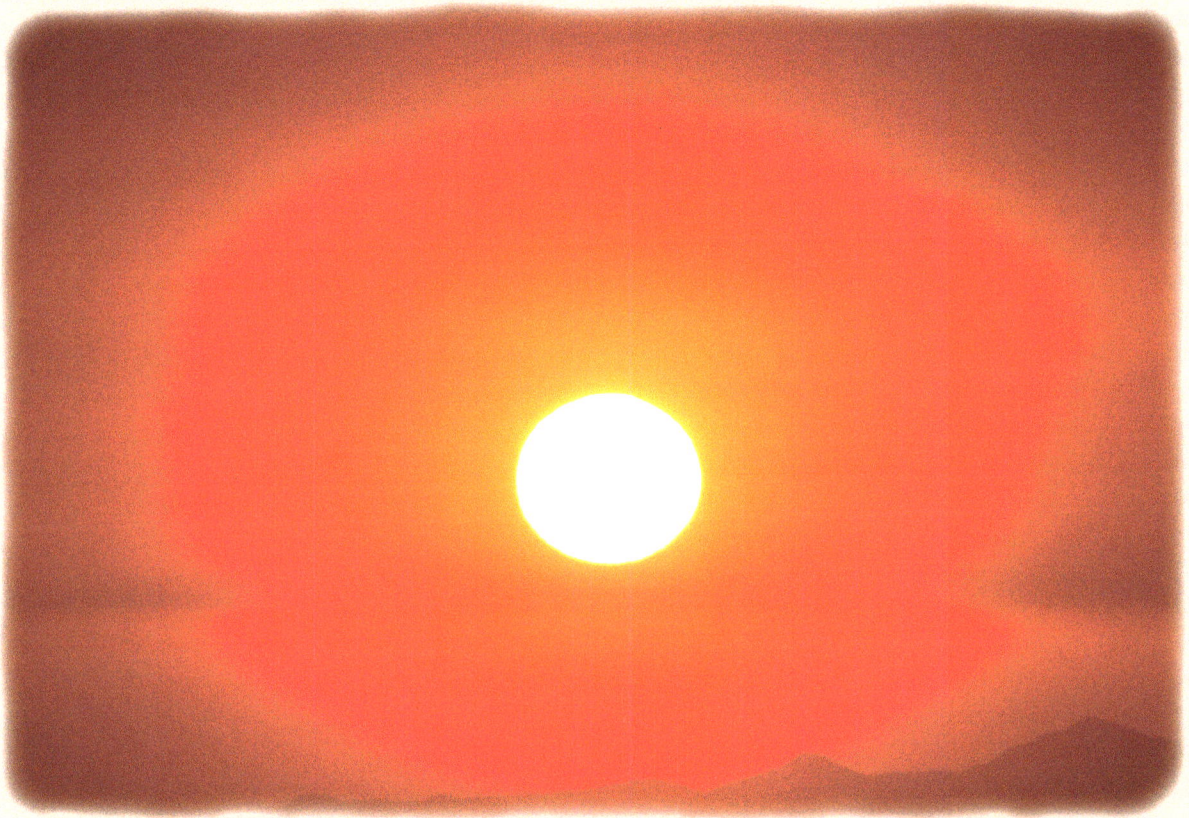

Photo by Luis Graterol on Unsplash

Light is a form of **energy**.

In **physics**, we say that energy gives something the ability to do **work**.

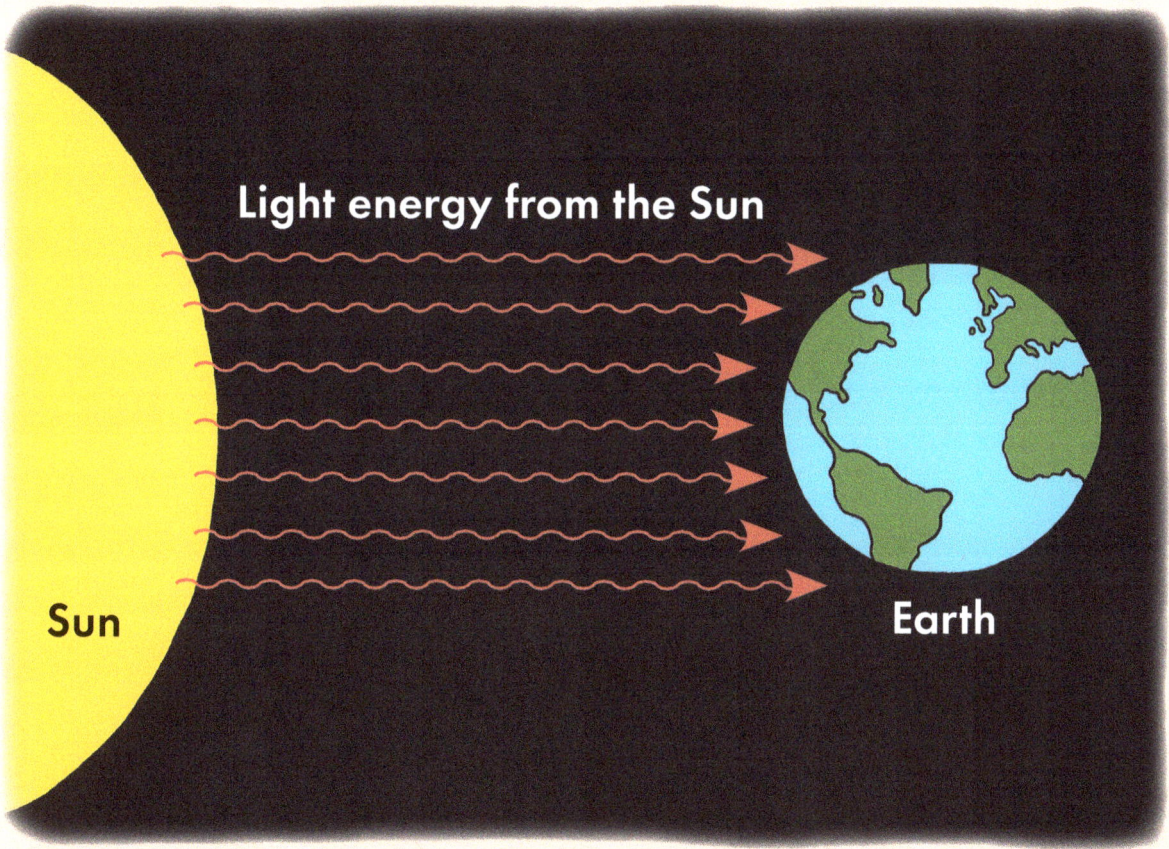

Light energy from the Sun

Sun

Earth

Visible light energy from the Sun is used by plants to make their own food in the process called **photosynthesis**.

Did you know plants use sunlight to make food?

Yes! And **we** use plants for food.

Light energy from the Sun gives our eyes the ability to see the world around us.

Light also comes from other sources.
We use flashlights to make light so
we can see when it is dark.

Light behaves in different ways.

Light bounces off objects that are **opaque**. The light bounces off an opaque object and goes to our eyes. This allows us to see objects.

Light goes through clear objects, such as window glass. Objects that light passes through are called **transparent**.

Light bounces off the objects behind window glass. The light then goes through the transparent glass to our eyes.

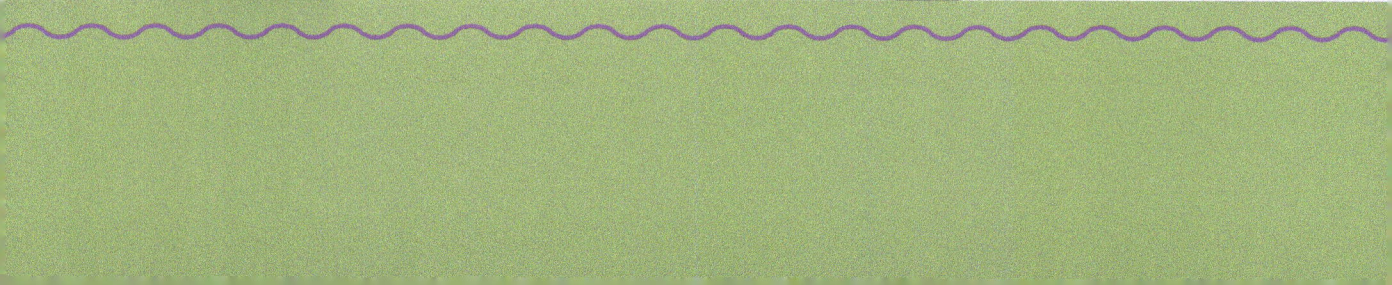

Some objects have part of the light bounce off them and the rest of the light pass through them. These objects are called **translucent**.

Stained glass is an example of a translucent object. From the side of the stained glass that is facing the source of light, you can see the part of the light that bounces off. From the other side of the stained glass you can see the part of the light that passes through.

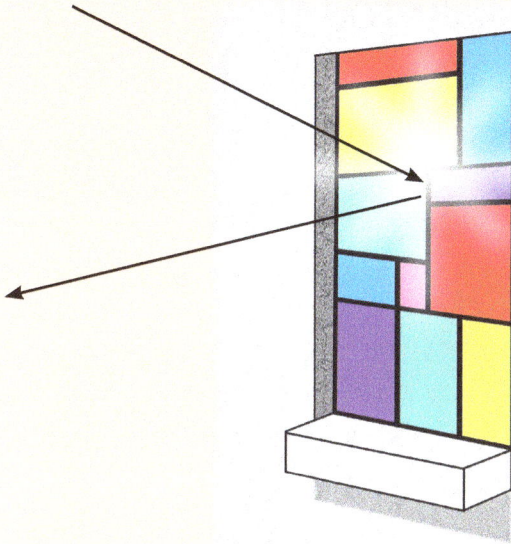

Some of the light bounces off the side of the stained glass that is facing the light source.

Some of the light passes through the stained glass.

Light energy from the Sun makes it possible for life to exist on Earth.

How to say science words

energy   (E-nuhr-jee)

light   (LIYT)

opaque   (oh-PAYK)

photosynthesis   (foh-toh-SIN-thuh-sis)

physics   (FIZ-iks)

sceince   (SIY-ens)

translucent   (trans-LOO-suhnt)

transparent   (trans-PER-uhnt)

visible   (VI-zuh-buhl)

work   (WERK)

What questions do you have about VISIBLE LIGHT?

Learn More Real Science!

Complete science curricula from Real Science-4-Kids

Focus On Series

Unit study for elementary and middle school levels

Chemistry
Biology
Physics
Geology
Astronomy

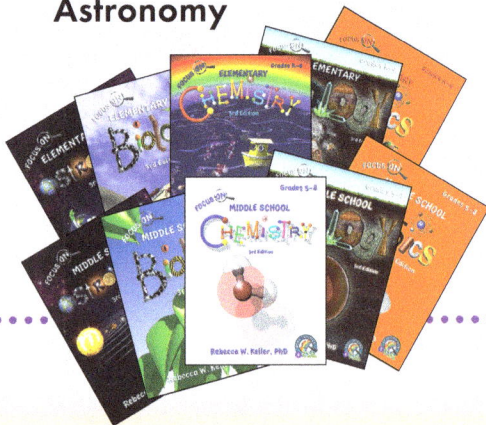

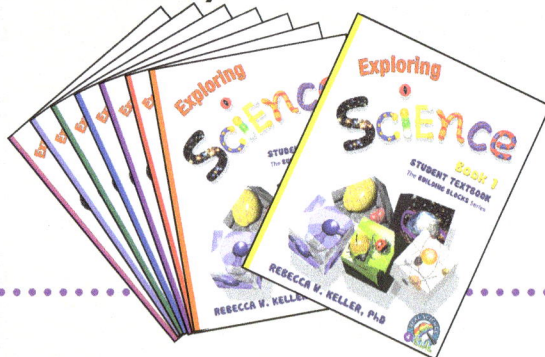

Exploring Science Series

Graded series for levels K–8. Each book contains 4 chapters of:

Chemistry
Biology
Physics
Geology
Astronomy

www.ingramcontent.com/pod-product-compliance
Lightning Source LLC
Chambersburg PA
CBHW040150200326
41520CB00028B/7555